FLORA OF TROPICAL EAST AFRICA

BREXIACEAE

B. Verdcourt

Trees or shrubs. Leaves coriaceous, simple, alternate or rarely subopposite or subverticillate, entire to spinously toothed; stipules present or absent. Flowers axillary, solitary or in cymes or false umbels. Sepals 4–6, imbricate or valvate, persistent or falling. Petals 4–6, imbricate and clawed or valvate, deciduous or persistent. Disk annular or joined with the staminodes and 5-lobed. Stamens 4–6, hypogynous to perigynous, free; anthers dithecous, opening lengthwise, usually large. Ovary superior, syncarpous, 4–7-locular with 2–numerous ovules in each locule; style 1, capitate, lobed or punctiform. Fruit a capsule, drupe or berry.

A small family of three genera and at a maximum estimate 10 species, frequently included in the *Escalloniaceae* but undoubtedly closely allied to the *Celastraceae*.

BREXIA

Thouars, Gen. Nov. Madag.: 20 (1806), *nom. conserv.*

Thomassetia Hemsl. in Hook., Ic. Pl. 28, t. 2736 (1902)

Mostly small trees or shrubs, glabrous. Leaves alternate, entire or toothed, sometimes spinously; stipules minute, very deciduous. Flowers in umbel-like cymes, sometimes on the old wood. Sepals imbricate, not persistent. Petals coriaceous, spreading, imbricate and twisted to right or left in the bud. Disk 5-lobed, the lobes alternating with the stamen-bases and each bearing 3–4(–6) stiff filaments (these lobes may be interpreted as clusters of staminodes). Stamens 4–5, slightly perigynous, inserted between the lobes of the disk; filaments slightly dilated near the base; anthers dehiscing inwards or sideways. Ovary elongate-ovoid, 5–10-angled, completely or imperfectly 5–7-locular, each locule with numerous ovules; style thick, simple, with 5–7-lobed stigma. Fruit at first woody, 5–7-locular or eventually becoming 1-locular, many-seeded, indehiscent. Seeds with sparse albumen; embryo as long as the seed.

A small genus, usually considered to be monotypic, but according to Perrier de la Bâthie containing 9 species, mainly confined to Madagascar, but extending to the Seychelles and the East African coast. Without a complete revision of the Madagascan material, which is mostly devoid of ripe fruits, it is difficult to assess the validity of the taxa involved. They may, at least in part, be more in the nature of races rather than distinct species. Certainly all the Seychelles material (*B. microcarpa* Tul.) has much smaller, mostly practically unribbed fruits and I would be prepared to accept it as subspecifically distinct from *B. madagascariensis**.

B. madagascariensis (*Lam.*) *Ker-Gawl.*, Bot. Reg. 9, t. 730 (1823); Thonner, Blütenpfl. Afr., t. 61 (1908); V.E. 1: 244, fig. 211 (1910); V.E. 3(1): 286 (1915); Thonner, Fl. Pl. Afr., t. 60 (1915); Engl. in E. & P. Pf.,

* The majority of writers have been prepared to treat *B. microcarpa* Tul. as a total synonym of *B. madagascariensis*, e.g. Hemsley in J.B. 54, Suppl. 2: 13 (1916) and Summerhayes in Trans. Linn. Soc. (Zool.) 19: 277 (1931).

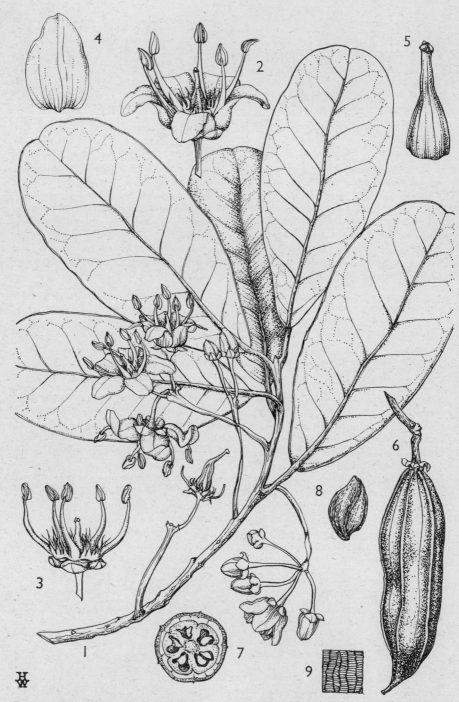

FIG. 1. *BREXIA MADAGASCARIENSIS*—1, flowering branch, × ⅔; 2, flower, × 1; 3, flower, with petals removed, × 1; 4, petal, × 2; 5, gynoecium, × 3; 6, fruit, × ⅔; 7, fruit, transverse section, × ⅔; 8, seed, × 3; 9, surface of seed, × 50. 1–5, from *Faulkner* 2669; 6, 8, 9, from *Faulkner* 2749; 7, from *Gomes e Sousa* 4426.

ed. 2, 18a: 186, fig. 104/A–F (1930); Perrier in Bull. Soc. Bot. Fr. 80: 202 (1933); U.O.P.Z.: 154 (1949); T.T.C.L.: 195 (1949). Type: Madagascar, *Commerson* (P, holo.)

Much-branched evergreen shrub or small tree, 2–6(–10) m. tall. Leaf-blades very variable in shape, narrowly oblong or linear-oblong to broadly obovate, with both sorts on the same plant (? not in the Flora area), the former on the young shoots, 3·5–35 cm. long, 2–7·6 cm. wide, broadly rounded or even retuse at the apex, rounded or cuneate at the base, ± entire, obscurely crenate or spinously toothed; petiole 1–2 cm. long. Inflorescences (1–)3–12(–17)-flowered; peduncles 1–9 cm. long, mostly flattened or even ± winged, 1·5–5 mm. wide, sometimes with a leaf-like bract ± 1 cm. long and wide at the apex; pedicels 0·6–1·8 cm. long; bracts scaly, small. Calyx mostly 0·8–1·2 cm. wide, the sepals ± 2·5 mm. long, 3·5–4 mm. wide, very rounded. Petals thick, greenish- or yellowish-white, elliptic-oblong, 1·2–1·7 cm. long, 0·9–1·2 cm. wide, obtuse. Ovary 5-angled. Fruit ovoid, oblong-fusiform or cylindrical, 4–10 cm. long, 1·9–3 cm. across, sometimes tapering, prominently 5-ribbed (in Flora area); all material examined has had very woody walls but the fruits are said to become pulpy at length and to be edible; they are capable of floating in the sea for several months without the seeds losing their viability. Seeds brown or blackish, irregularly compressed-ellipsoid, 4·5–7·5 mm. long, 3–3·5 mm. wide, keeled, minutely rugulose in ridges. Fig. 1.

TANGANYIKA. Uzaramo District: Kisiju, Sept. 1953, *Semsei* 1366!; Rufiji District: Mafia I., Jibondo I., 25 Sept. 1937, *Greenway* 5312! & Mafia I., Ras Mbisi, 27 July 1932, *Schlieben* 2598!
ZANZIBAR. Zanzibar I., Masingini Ridge, 1 Feb. 1929, *Greenway* 1295! & Dole Woods, 28 Aug. 1950, *R. O. Williams* 65! & Kidichi, 22 July 1960, *Faulkner* 2669! & 27 Jan. 1961, *Faulkner* 2749!
DISTR. **T6**; **Z**; Mozambique, Madagascar, Comoro Is., ? Seychelles
HAB. Coastal evergreen bushland on coral or coarse rocky ground, also at the edges of saline water swamp forest, mangrove swamps and on eroded ridges near the sea; 0–10 m.

SYN. *Venana madagascariensis* Lam., Illustr. Gen. 2: 99 (1797), t. 131 (1792); Poir., Encycl. Méth. 8: 450 (1808)
       Brexia madagascariensis (Lam.) Ker-Gawl. var. *mossambicensis* Oliv., F.T.A. 2: 388 (1871). Type: Mozambique, mouth of the Zambezi, *Kirk* (K, holo.!)

DISTR. (of species as a whole). If the Seychelles populations are considered to form a subspecies rather than a species, as above with the addition of the Seychelles Archipelago. Perrier de la Bâthie records B. *madagascariensis* sensu stricto from the Seychelles.

VARIATION. In East Africa no specimens have been seen bearing long narrow spinous leaves; all those seen have obovate leaves 3·5–16 cm. long, 2·9–7·6 cm. wide, with entire or slightly crenate margins and can be exactly matched with Madagascan material. There may, however, be a tendency for the mainland material to have larger fruits and fewer-flowered inflorescences. If further comparisons indicate these are constant differences, then Oliver's varietal designation can be upheld.

INDEX TO BREXIACEAE

FLORA OF TROPICAL EAST AFRICA

BASELLACEAE

B. Verdcourt

Subsucculent, glabrous, twining herbs with slender stems and alternate, entire, petiolate or rarely sessile, exstipulate leaves. Flowers regular, hermaphrodite or unisexual, in spikes, racemes or panicles; bracts small; bracteoles 2–4, often 2 adnate to the base of the perianth, sometimes wing-like. Perianth 5-lobed; lobes imbricate, sometimes coloured, united into a tube below or almost free, persistent. Stamens 5, opposite to the perianth-lobes, inserted at their base; filaments free, short; anthers versatile, variously dehiscing. Ovary superior, 1-locular; ovule solitary, basal, shortly stalked, campylotropous; style terminal, simple or 3-fid, or 3 free styles. Fruit indehiscent, surrounded by the persistent often fleshy perianth or winged bracteoles. Seeds solitary, almost spherical; endosperm copious or almost absent, surrounded by the spirally twisted or semi-annular embryo.

A small tropical family of four or five genera, only one of which occurs in East Africa.

BASELLA

L., Sp. Pl.: 272 (1753) & Gen. Pl., ed. 5: 133 (1754); Hook. f., G. P. 3: 76 (1880); Perrier de la Bâthie in Not. Syst. 14: 53–56 (1950)

Twining subsucculent herbs with long much-branched stems. Leaves alternate; lamina ovate, entire. Flowers ⚲, sessile, white or coloured, in axillary spikes or panicles of spikes. Bracteoles 2, united into a 2-lipped cup adnate to the perianth. Perianth urceolate or wide open, fleshy; lobes 5, short, incurved, or almost free and spreading. Stamens 5, inserted near the top of the perianth-tube. Ovary ovoid, free; styles 3, or 1 deeply trifid or obsolete; stigmas 3, linear. Fruit membranous, one-seeded. Seed globose; embryo spiral; endosperm scanty.

A small genus of about five species, three very distinctive ones endemic in Madagascar, one endemic to eastern Africa and a polymorphic one widespread in Africa and Asia and also cultivated (? always) in America; the latter is widely cultivated as a spinach-like vegetable.

The actual interpretation of the flower-structure is not clear; the bracteoles could be treated as sepals adnate to a corolla but I have preferred to call the main part a perianth. I suspect it is calycine as in related families.

Inflorescences mostly simple spikes, sometimes branched; petioles of leaves up to 6·5 cm. long; perianth with conspicuous tube, not opening in anthesis; united part of styles developed . . 1. *B. alba*

Inflorescence consisting of a panicle of spikes; petioles of leaves ± 5 mm. long; perianth lobed almost to the base, wide open in anthesis; stigmas sessile . . . . 2. *B. paniculata*

1. **B. alba** *L.*, Sp. Pl.: 272 (1753); Roxb., Fl. Ind. 2: 104 (1832); Moq. in DC., Prodr. 13 (2): 223 (1849); Volkens in E. & P. Pf. 3(1a): 126, fig. 73/A–F (1893); Bak. & C.B. Cl. in F.T.A. 6(1): 94 (1909); F.D.O.-A. 2: 272 (1938); F.P.N.A. 1: 155 (1948); Hauman in F.C.B. 2: 129 (1951); U.O.P.Z.: 141, fig. (1949); F.W.T.A., ed. 2, 1: 155, fig. 57 (1954); van Steenis in Fl. Males., ser. 1, 5(3): 300, fig. 1 (1957). Type: *Basella flore albo, foliis & caulibus viridibus* in Thran, Hort. Carolsruh. 10, n. 100 (1747)

Glabrous annual or shortly lived perennial, succulent tangled twiner; stems much branched, 2–10 m. long, sometimes almost leafless, greenish or reddish. Leaf-lamina ovate to suborbicular, (2–)5–15 cm. long, (1·25–)5–13·5 cm. broad, acute or acuminate (less commonly obtuse), usually widely cordate at the base; lateral nerves 4–5 on either side; petiole (1–)2·5–6·5 cm. long. Flowers white, rose or purplish, (3–)4–5 mm. long, in long-peduncled spikes, 2·5–15(–25) cm. long, usually unbranched (in African specimens at least) but branched in some cultivated forms. Perianth fleshy, urceolate, somewhat saccate at the base; lobes short, ovate, about one-third the length of the tube, not opening. Fruits ± 0·5 cm. in diameter (4–7 × 5–10 mm. according to van Steenis), red, white or black; surface crinkly in the dry state. Fig. 1/1–10.

UGANDA. Kigezi District: Kachwekano Farm, July 1949, *Purseglove* 3034!; Mengo District: Semunya Forest, 13 km. NW. of Entebbe, 13 June 1950, *Dawkins* 590! & Entebbe, Oct. 1931, *Eggeling* 49 in *F.H.* 218!
KENYA. Northern Frontier Province: southern slopes of Mt. Kulal, Aug. 1957, *J. Adamson* 669!; Fort Hall/Machakos Districts: banks of R. Athi near Donyo Sabuk, 27 Aug. 1956, *Verdcourt* 1567!; Masai District: Mara Masai, Telek R., 26 Feb. 1950, *Kirrika* in *C.M.* 17482!
TANGANYIKA. Arusha District: Themi [Temi] R., 8 Sept. 1943, *Lindeman* 822!; Moshi District: Lyamungu, 6 Oct. 1943, *Wallace* 1112!; Lushoto District: W. Usambara Mts., Mkuzi to Kwai road, 21 May 1953, *Drummond & Hemsley* 2676!
ZANZIBAR. Zanzibar I., Residency Garden, 21 July 1962, *Faulkner*! (cult.) & without locality, *Oxtoby*!
DISTR. U2, 4; K1, 3, 4, 6; T1–4, 6, 7; Z; Madeira, West Africa to Cameroun Republic, S. Tomé, Congo Republic, Sudan Republic, Ethiopia, Rwanda Republic, Mozambique, Malawi, Zambia and Angola but rare in central Africa; Asia to China, Japan, Philippines, Borneo, Fiji and Hawaii, also in West Indies, Brazil and Guiana; almost certainly indigenous in Africa.
HAB. In thickets, forest edges, margins of cultivated land and swampy ground, frequently by rivers or streams; 0 (cultivated)–2450 m.

SYN. *B. rubra* L., Sp. Pl.: 272 (1753); Hook. f. in Fl. Brit. India 5: 20 (1886); F.P.S. 1: 123, fig. 74 (1950); E.P.A.: 92 (1953). Type: drawing of fruiting plant in *Herb. Hermann* 5, t. 207 (BM, lecto.!)
B. cordifolia Lam., Encycl. 1: 382 (1785); Moq. in DC., Prodr. 13 (2): 223 (1849). Type: Malabar (cultivated): illustration of *Basella* in Rheede, Hort. Malab., t. 24 (1688)
For more complete synonymy see the above F.T.A. reference.

NOTE. Some cultivated plants, e.g. *Greenway* 6167! from Tanganyika, Amani, probably originally from Hong Kong, have much thicker stems and larger leaves with relatively thick petioles; the inflorescences are sometimes branched. They closely resemble the cultivated plant depicted by Rheede and if a varietal name is required for this ennobled plant, it could be based on Lamarck's epithet, *cordifolia*.

2. **B. paniculata** *Volkens* in E.J. 38: 81 (1905); Bak. & C.B. Cl. in F.T.A. 6 (1): 94 (1909); F.D.O.-A. 2: 272 (1938); Verdc. in K.B. 17: 496 (1964). Type: Tanganyika, E. Pare Mts., Kihurio, *Engler* 1515 (B, holo. †)

Glabrous subsucculent twiner, often almost leafless; stems rather thick, furrowed, rough, pale brown. Leaf-blade ovate, 1·8–3·5 cm. long, 1–2·5 cm. wide, acute, truncate to cuneate at the base; petiole 2–5 mm. long. Flowers small, greenish or white, in spikes, with the spikes arranged in much branched rather dense panicles up to 8 cm. long. Perianth 5-lobed almost to the base,

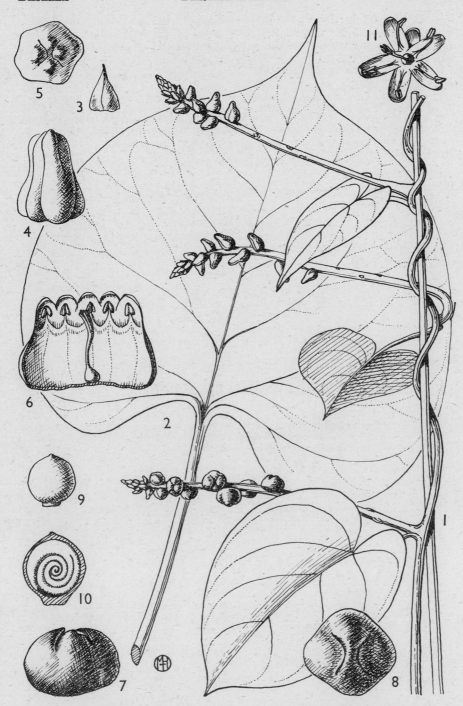

FIG. 1. *BASELLA ALBA*—**1**, habit, × 1; **2**, leaf, × 1; **3**, bracteole, × 4; **4**, flower-bud, side view, × 4; **5**, same, front view, × 4; **6**, flower, opened out, × 4; **7**, fruit, side view, × 3; **8**, same, front view, × 3; **9**, seed, × 3; **10**, vertical section of same, × 4. *B. PANICULATA*—**11**, flower, × 4. 1, from *Verdcourt* 1567; 2, from *Fries* 846; 3–6, from *Milne-Redhead & Taylor* 11402; 7–10, from *Faulkner* (Zanzibar, Residency Garden, 1962); 11, from *Milne-Redhead & Taylor* 7244.

opening completely; lobes oval or suborbicular, 2–2·5 mm. long, 2 mm. wide, rounded at apex. Fruits globose, 3·5–4 mm. in diameter, pale brown (red *in vivo* ?), shining, with 6 prominent longitudinal ribs; receptacles swollen and fleshy. Fig. 1/11, p. 3.

KENYA. Teita District: Kasigao, Sept.–Oct. 1938, *Joanna* in *C.M.* 8878 ! ; Tana River District: 24 km. S. of Garsen on Malindi road, 10 Nov. 1957, *Greenway* 9491 !
TANGANYIKA. Pare District: Pare Hills, Makania, 13 Jan. 1945, *Bally* 4209 ! & 23 Nov. 1945, *Gillman* in *Herb. Amani* 9872 ! & Kisiwani, 7 Nov. 1955, *Milne-Redhead & Taylor* 7244 ! ; Pangani District: Mikocheni, 29 Nov. 1909, *Kränzlin* in *Herb. Amani* 2963 !
DISTR. **K7**; **T3**; also Mozambique and the Transvaal
HAB. Dry evergreen bushland and dry evergreen forest, often at the edges, on sandy loam or rocky ground; 60–1350 m.

NOTE. *B. paniculata* is very different in flower-structure from *B. alba* L., but they are best considered congeneric.

INDEX TO BASELLACEAE